I0796962
L'EXPOSITION
DE
PARIS 1900
Exposition Universelle de 1900

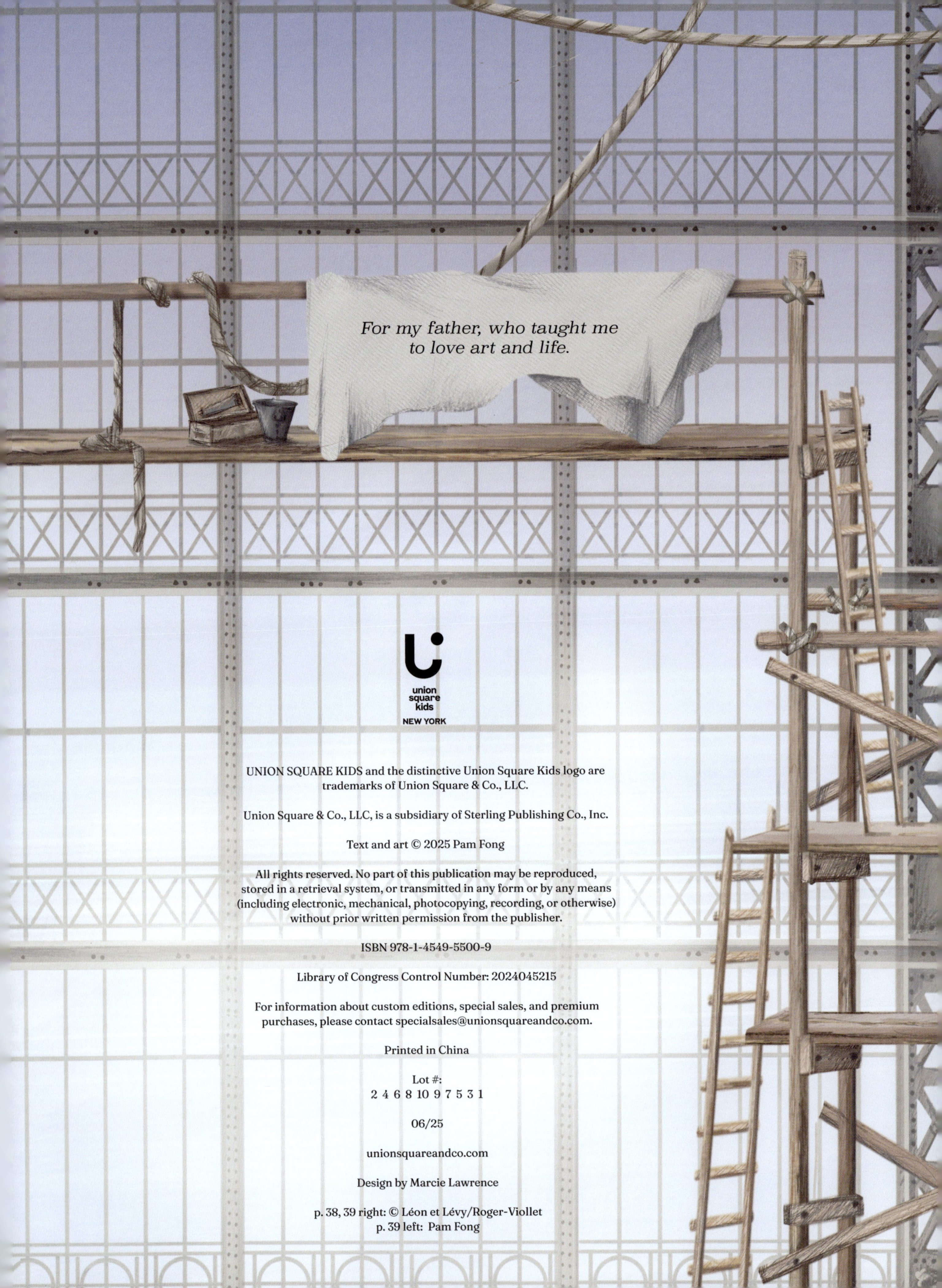

For my father, who taught me to love art and life.

UNION SQUARE KIDS and the distinctive Union Square Kids logo are trademarks of Union Square & Co., LLC.

Union Square & Co., LLC, is a subsidiary of Sterling Publishing Co., Inc.

Text and art © 2025 Pam Fong

All rights reserved. No part of this publication may be reproduced, stored in a retrieval system, or transmitted in any form or by any means (including electronic, mechanical, photocopying, recording, or otherwise) without prior written permission from the publisher.

ISBN 978-1-4549-5500-9

Library of Congress Control Number: 2024045215

For information about custom editions, special sales, and premium purchases, please contact specialsales@unionsquareandco.com.

Printed in China

Lot #:
2 4 6 8 10 9 7 5 3 1

06/25

unionsquareandco.com

Design by Marcie Lawrence

p. 38, 39 right: © Léon et Lévy/Roger-Viollet
p. 39 left: Pam Fong

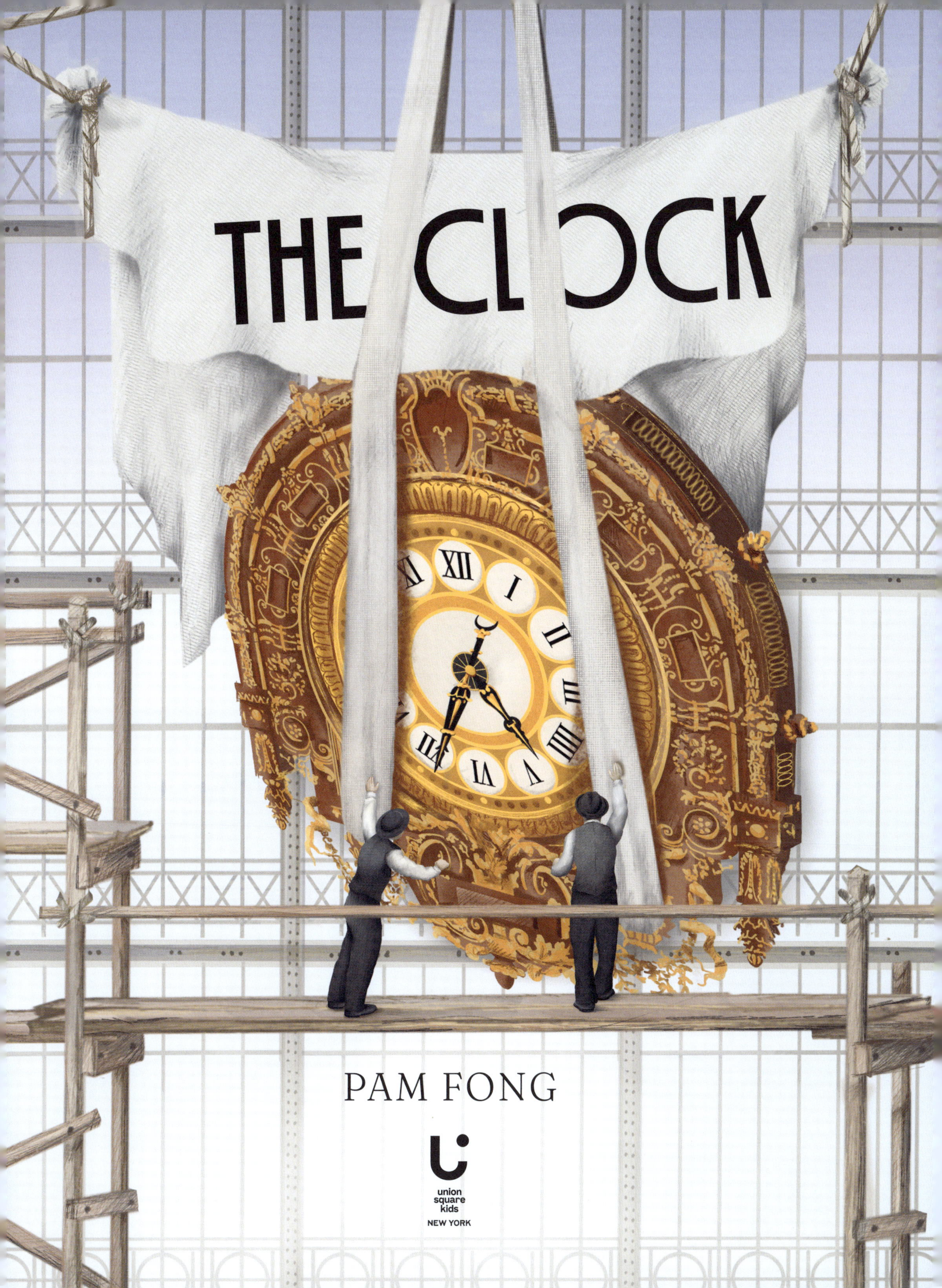

THE CLOCK

PAM FONG

union square kids
NEW YORK

BILLETTERIE
Trains Urbains & Ailleurs
TRAINS DE PLAISIR
PARIS-AUSTERLITZ

In another time, in a growing city,
there was an important clock.

The clock ***ticked*** to control the crisscrossing of trains.

The clock ***tocked*** to signal
new opportunities.

Ticktock, ticktock. The clock was there to welcome the world.

BONNE ANNÉE 1939

Minute by minute, day after day, year after year
the clock kept a station humming,
and the world moving.

But there came a time when that
fast-changing city outgrew the station.

The clock ***ticked*** as the crowds disappeared.

The clock ***tocked*** as the trains stopped.

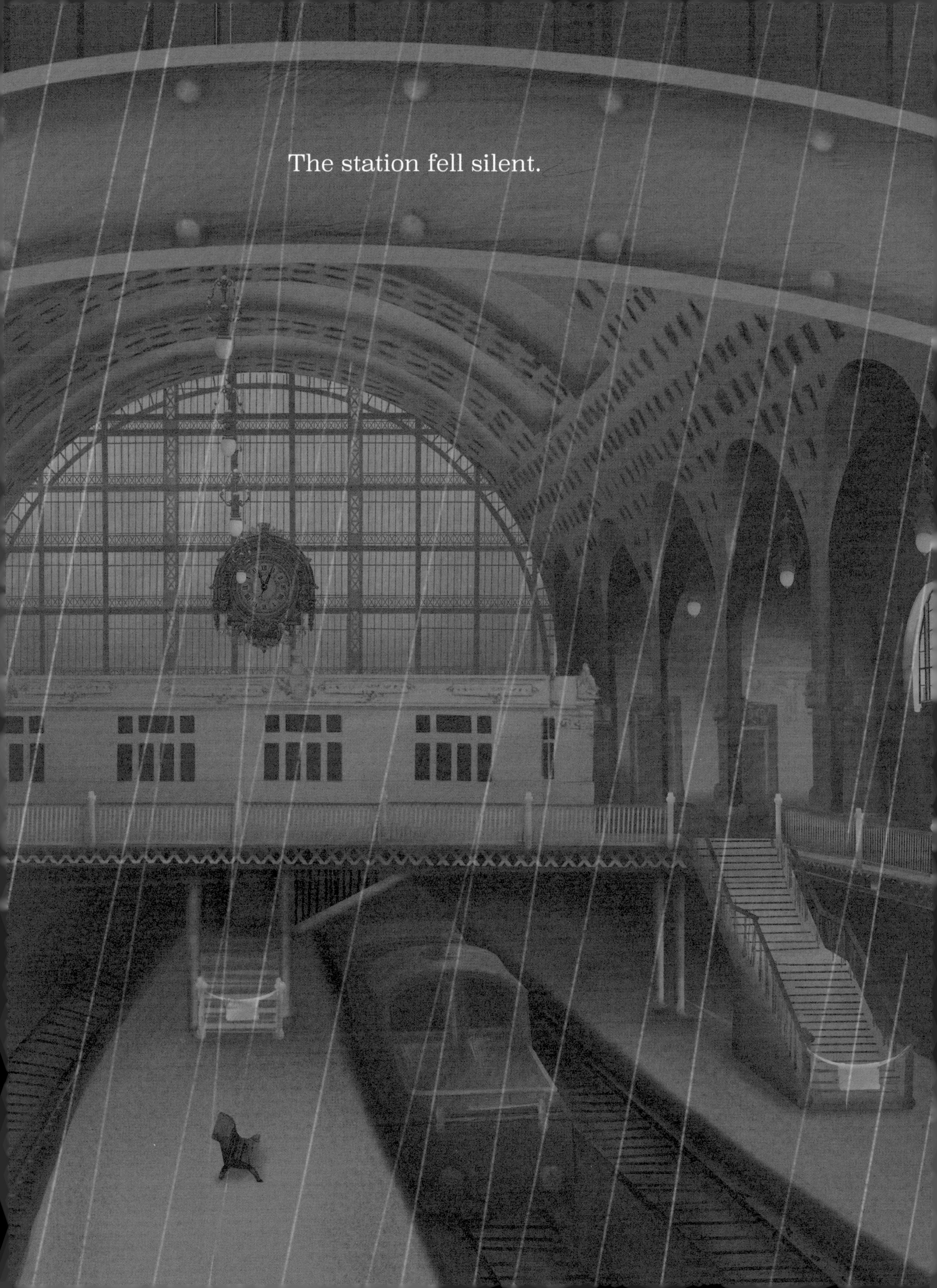

The station fell silent.

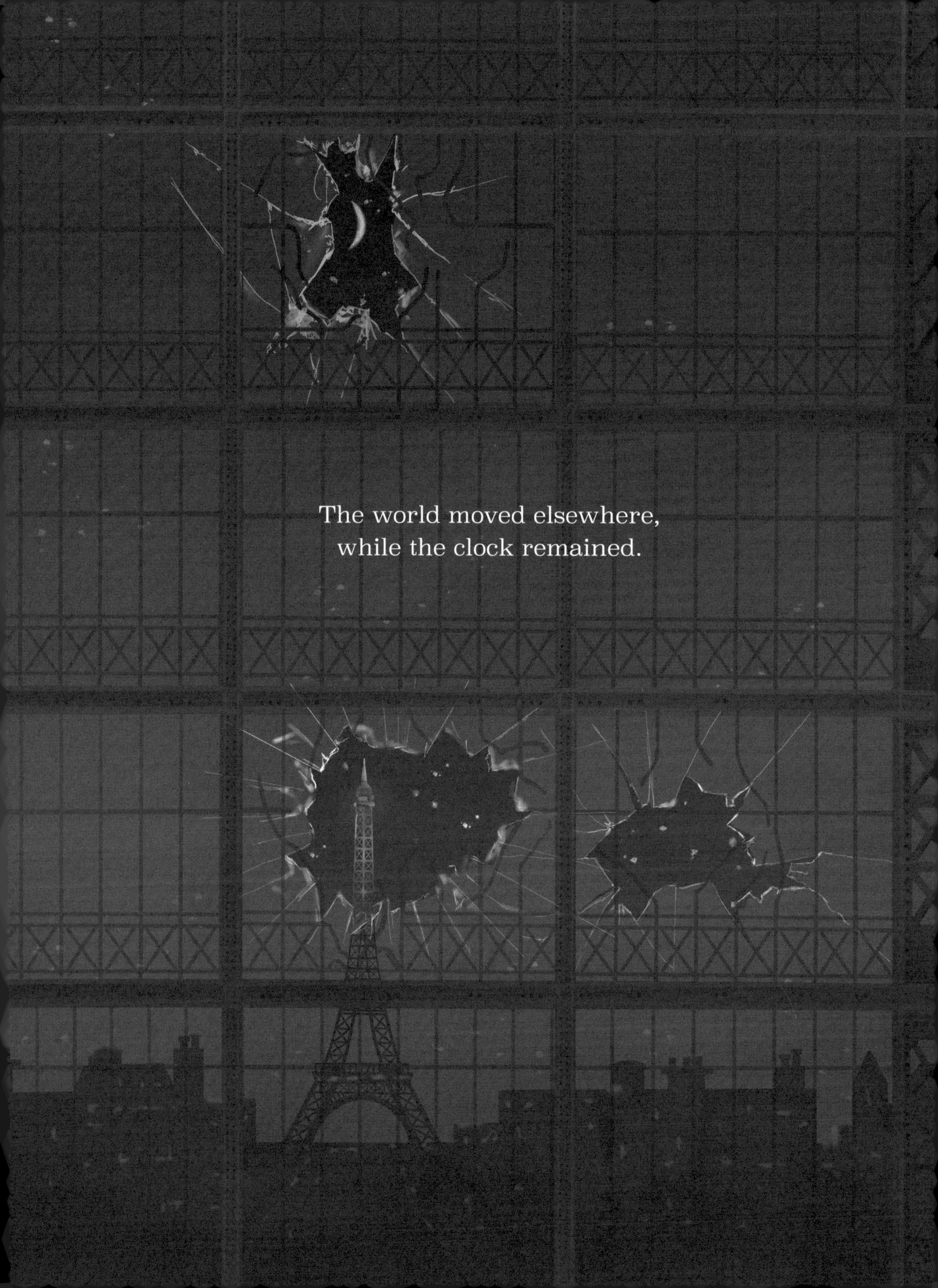

The world moved elsewhere,
while the clock remained.

Ticktock. Ticktock.
Until its final *ticktock*.

Days turned into years.
Years into decades.

The clock was abandoned.

But it was not forgotten.

There were those who remembered the clock.
Who loved the clock for its beauty.
Who cherished the clock for its memories.
Who admired the clock for withstanding time.

And who saw an important
role in its future.

CLA
MONET

Today, in that famous city, there is a
beloved treasure. A clock that once
again welcomes the world.

MORISO

Ticktock, ticktock.
Minute by minute, day after day, for years to come. This clock keeps a world-class art museum humming . . .

. . . and stops the world
from moving.

View of the train platform in the central terminal of the Gare d'Orsay in the early 1900s.

Originally built as a train station, the Gare d'Orsay (Orsay Station) was the vision of architect Victor Laloux. It was constructed in two short years and was inaugurated in July 1900 in conjunction with the World Fair in Paris.

The Gare d'Orsay was an architectural achievement of French engineering. Prior to its construction, buildings were made of stone or brick. The Gare d'Orsay was constructed with 12,000 tons of steel and 376,737 square feet (the equivalent of more than six football fields) of glass, allowing the interior to flood with natural light.

The advancements of the Gare d'Orsay were not only limited to its architecture. The trains were the first urban rail system in the world to run solely on electricity, replacing the steam engine. The Gare d'Orsay stood in stark contrast to the dark, soot-filled interiors that characterized prior train stations.

Prominently displayed in the main terminal was a clock, measuring 15 feet in diameter, also designed by Laloux and powered by electricity. In a time when few people but the rich owned watches, the majority of travelers relied on the central clock. With the timed arrivals and departures of hundreds of trains daily, the clock was essential to the station.

Despite all this, four decades after its inauguration, the station was abandoned. The short platforms could no longer accommodate the expanding length of the modern trains needed to service the growing city. In 1970, permission was granted to demolish the station and repurpose the land for new construction. However, the Minister for Cultural Affairs intervened to stop the demolition. At the same time, three of the city's largest museums were struggling with overflowing art collections. It was decided that the Gare d'Orsay would be transformed into a museum to house the priceless works.

Automated chutes delivered luggage from the first floor down to the trains.

The station transformed into the Musée d'Orsay while retaining much of its original character.

ACT architecture group presented the winning renovation design, and work began in 1979. Italian architect, Gae Aulenti, was entrusted with the task of designing the interiors, including installation of the artworks that took over six months. On December 9, 1986, the Musée d'Orsay (Orsay Museum) opened to the public.

Nearly 4 million visitors each year come to view the collection and experience the architecture. It is home to some of the greatest Western artistic works ever created, including Impressionism and Post-Impressionism masterpieces by Claude Monet, Vincent Van Gogh, Paul Cézanne, Camille Pissarro, and Edouard Manet. Among these treasures, one is still faithfully keeping time from high above: the clock.

ALFRED SISLEY
BERTHE MORISOT
Georges Seurat
Cézanne
Du 30 septembre 1995 au 07 janvier 1996